I0483768

Los Alamos National Laboratory

&

Lawrence Livermore National Laboratory

Plutonium Sustainment Monthly Program Report

April 2012

Executive Summary

In April of 2012 the Plutonium Sustainment program at LANL completed or addressed the following high-level activities:

1) Construction activities on the Low Voltage Electron Beam (LVEB) Welder and Analytical Chemistry gloveboxes have started, successfully closing out MRT L2 Milestone 4195.
2) Completed installation of the Surface Preparation unit in PF4. The unit will be functionally tested in May.
3) Completed casting of the first two near net shape W87 hemishells.
4) Delivered Revision 3 of the Plutonium Sustainment and Manufacturing Study.
5) Completed machining and inspecting the first set of Pollux parts on Precitech 1.
6) Attended the Pu Sustainment Second Quarter Mid-Year Review and the Pit Production Integrated Project Team meeting in Livermore, providing technical updates across the Plutonium Sustainment portfolio.
7) Analyzed the program impact of the LANL Voluntary Separation Program. The analysis determined an impact of 24.6 FTEs and a budget impact of approximately $3.2 Million Dollars of pull back.
8) The Power Supply Assembly Area project received a LANL Pollution Prevention "Gold" award in the category of Cradle-to-Cradle – Cleanouts. A gold award is presented to the number one project in its pollution prevention category.

There are currently no major issues associated with achieving MRT L2 Milestones 4195-4198 or the relevant PBIs associated with Plutonium Sustainment.

There are no budget issues associated with FY12 final budget guidance. Table 1 identifies all Baseline Change Requests (BCRs) that were initiated, in process, or completed during the month.

The earned value metrics overall for LANL are within acceptable thresholds, so no high-level recovery plan is required.

Each of the 5 major LANL WBS elements is discussed in detail below with a section at the end that is provided by LLNL.

Table 1: April Baseline Change Requests

BCR #	BCR Description	Date Assigned	Date Approved	Impact	Cost Impact	Change Level
PSM-12-027	Addition of KK05 0001 0000 Cost String to Program Mgmt. WP	3/27/12	3/27/12	N/A	N/A	4
PSM-12-028	JT29 Alignment	3/28/12	3/29/12	N/A	N/A	4
PSM-12-030	Electrical Install - In scope Scope changes	4/20/12		Scope	N/A	
PSM-12-031	Mechanical Install - In scope Scope changes	4/20/12		Scope	N/A	
PSM-12-032	Civil/Structural/Architectural - In scope Scope changes	4/20/12		Scope	N/A	
PSM-12-033	Demolition - Mechanical	4/20/12		Scope	N/A	
PSM-12-034	Scope Modification for Pu Separation & Granule Production	5/1/12		Scope	N/A	3
PSM-12-035	Scope Modification for Heat Source, Fabrication & Inspection	5/1/12	5/9/12	Scope	N/A	3

Earned Value Metrics

PERFORMANCE S-CURVE
Pu Sustainment

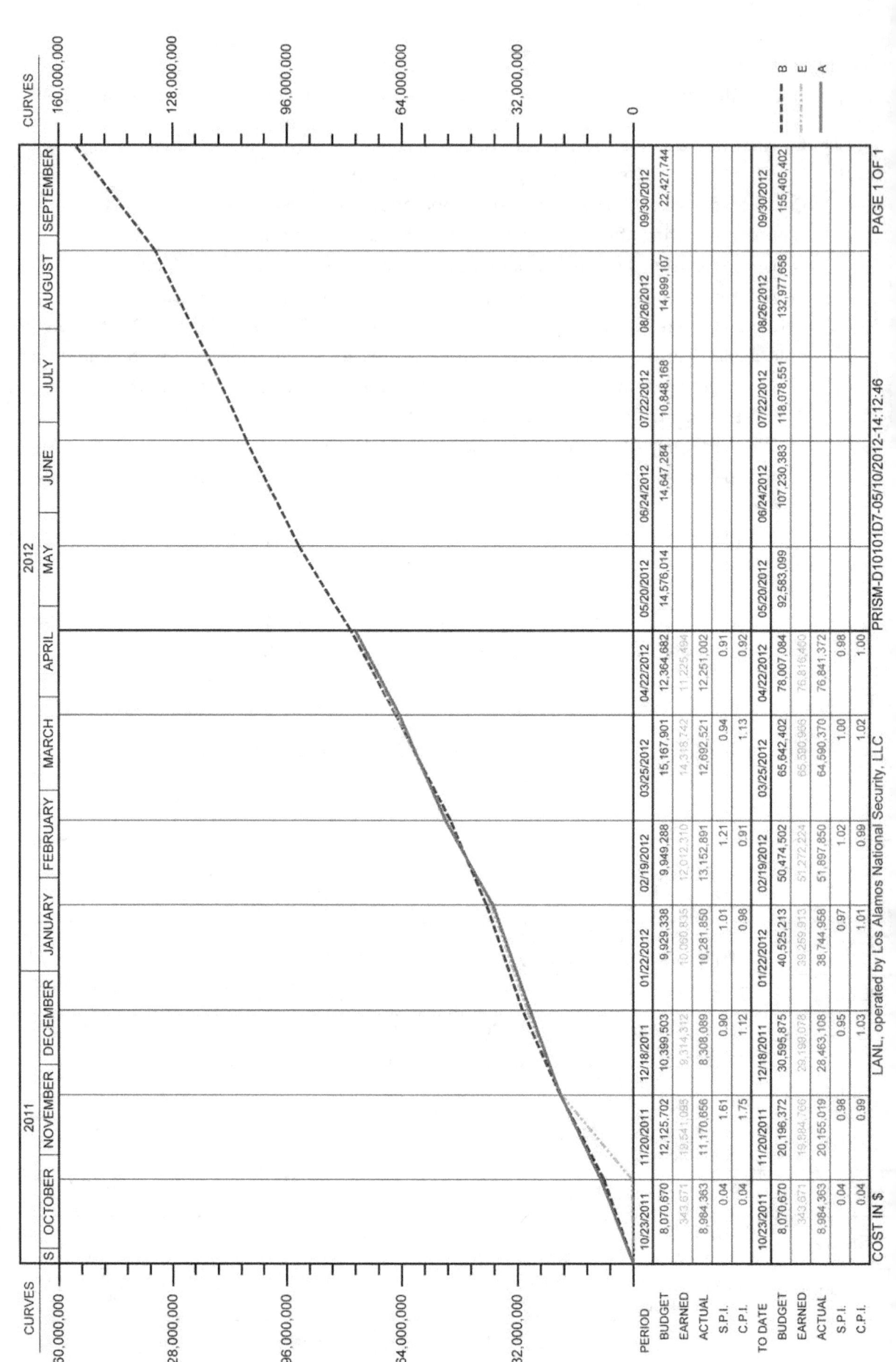

PERIOD		2011			2012								
	S	OCTOBER	NOVEMBER	DECEMBER	JANUARY	FEBRUARY	MARCH	APRIL	MAY	JUNE	JULY	AUGUST	SEPTEMBER
		10/23/2011	11/20/2011	12/18/2011	01/22/2012	02/19/2012	03/25/2012	04/22/2012	05/20/2012	06/24/2012	07/22/2012	08/26/2012	09/30/2012
BUDGET		8,070,670	12,125,702	10,399,503	9,929,338	9,949,288	15,167,901	12,364,682	14,576,014	14,647,284	10,848,168	14,899,107	22,427,744
EARNED		343,671	18,541,095	9,314,312	10,060,835	12,012,310	14,315,742	11,225,494					
ACTUAL		8,984,363	11,170,656	8,308,089	10,281,850	13,152,891	12,692,521	12,251,002					
S.P.I.		0.04	1.61	0.90	1.01	1.21	0.94	0.91					
C.P.I.		0.04	1.75	1.12	0.98	0.91	1.13	0.92					
TO DATE		10/23/2011	11/20/2011	12/18/2011	01/22/2012	02/19/2012	03/25/2012	04/22/2012	05/20/2012	06/24/2012	07/22/2012	08/26/2012	09/30/2012
BUDGET		8,070,670	20,196,372	30,595,875	40,525,213	50,474,502	65,642,402	78,007,084	92,583,099	107,230,383	118,078,551	132,977,658	155,405,402
EARNED		343,671	19,894,766	29,199,078	39,259,913	51,272,224	65,590,966	76,816,460					
ACTUAL		8,984,363	20,155,019	28,463,108	38,744,958	51,897,850	64,590,370	76,841,372					
S.P.I.		0.04	0.98	0.95	0.97	1.02	1.00	0.98					
C.P.I.		0.04	0.99	1.03	1.01	0.99	1.02	1.00					

COST IN $ LANL, operated by Los Alamos National Security, LLC   PRISM-D10101D7-05/10/2012-14:12:46   PAGE 1 OF 1

CURVES: B — — — — E ········· A ——————

PERFORMANCE S-CURVE
Pu Sustainment W/O Facility Infrastructure

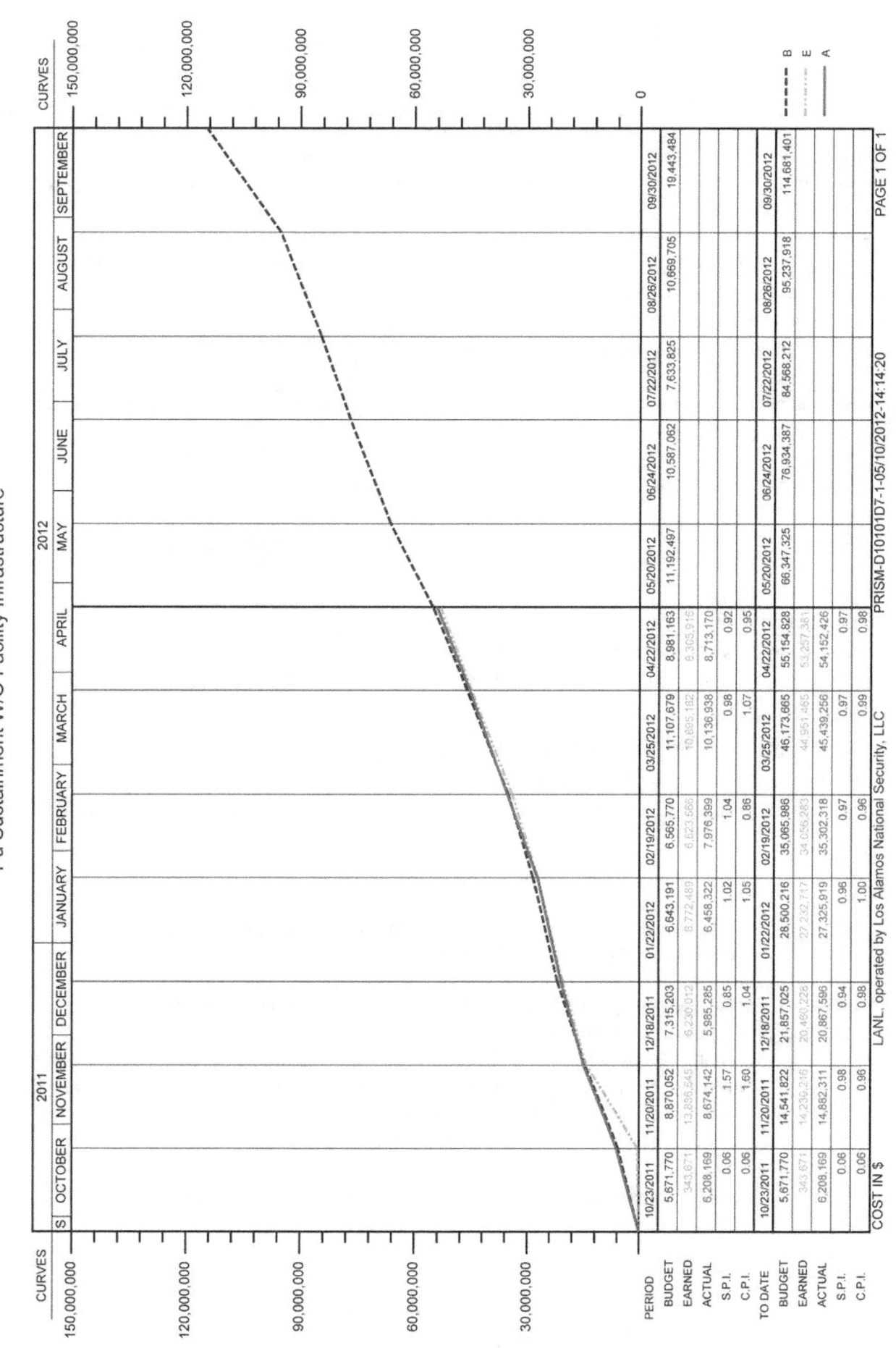

	2011			2012								
	OCTOBER	NOVEMBER	DECEMBER	JANUARY	FEBRUARY	MARCH	APRIL	MAY	JUNE	JULY	AUGUST	SEPTEMBER
PERIOD	10/23/2011	11/20/2011	12/18/2011	01/22/2012	02/19/2012	03/25/2012	04/22/2012	05/20/2012	06/24/2012	07/22/2012	08/26/2012	09/30/2012
BUDGET	5,671.770	8,870.052	7,315.203	6,643.191	6,565.770	11,107.679	8,981.163	11,192.497	10,587.062	7,633.825	10,669.705	19,443.484
EARNED	343.671	13,836.545	6,230.012	6,772.489	6,623.566	10,895.182	8,365.916					
ACTUAL	6,208.169	8,674.142	5,985.285	6,458.322	7,976.399	10,136.938	8,713.170					
S.P.I.	0.06	1.57	0.85	1.02	1.04	0.98	0.92					
C.P.I.	0.06	1.60	1.04	1.05	0.86	1.07	0.95					
TO DATE	10/23/2011	11/20/2011	12/18/2011	01/22/2012	02/19/2012	03/25/2012	04/22/2012	05/20/2012	06/24/2012	07/22/2012	08/26/2012	09/30/2012
BUDGET	5,671.770	14,541.822	21,857.025	28,500.216	35,065.986	46,173.665	55,154.828	66,347.325	76,934.387	84,568.212	95,237.918	114,681.401
EARNED	343.671	14,230.216	20,480.228	27,252.717	34,036.283	44,951.465	53,257.381					
ACTUAL	6,208.169	14,882.311	20,867.596	27,325.919	35,302.318	45,439.256	54,152.426					
S.P.I.	0.06	0.98	0.94	0.96	0.97	0.97	0.97					
C.P.I.	0.06	0.96	0.98	1.00	0.96	0.99	0.98					

COST IN $

LANL, operated by Los Alamos National Security, LLC

PRISM-D10101D7-1-05/10/2012-14:14:20

CURVES: B, E, A

1.0 LANL - Pit Development

MRT 4196	GREEN	*Perform Activities to Sustain Base Pit Material Processing and Fabrication Capability*
1. Complete FY12 LANL tasks per PEP and W87 Legacy IPT necessary to support FY2013 delivery of 1 pit EDU 2. Continue installation and upgrade to pit capability equipment per PEP 3. Continue supporting W87 IPT and deliver Production Strategy and TRL/MRL assessment by September 30, 2012		

L3 Milestone	Baseline Date	RGYB	Status/Comments
Non-nuclear components necessary for an EDU build are available for use.	July 13, 2012	GREEN	
Suitable nuclear material is available as ER	September 21, 2012	GREEN	
Laser Welding documentation approved • Technical Specifications & Requirements • Functional & Operations Requirements • Cold System Test Plan	September 21, 2012	GREEN	A BCR addressing scope and staffing issues has been approved.
Electron Beam Welder Corrective Maintenance Complete	June 29, 2012	GREEN	
Preparation Unit Installed	August 14, 2012	GREEN	
Production Strategy and TRL/MRL assessment	September 30, 2012	GREEN	

1.1 Technical Progress

- Completed the first two W87 near net shape castings.
- Continued machining and inspection operations for Transition Build 2.
- Completed Build 62 LASO customer product acceptance process and the build is ready for Diamond Stamping.

1.2 Equipment

- Laser Welder
    - Drafted portions the Technical Specifications and Requirements and started preparation of the Functional and Operational Requirements.
    - Resolved Software Quality Assurance issues associated with equipment.
    - Evaluated the options for replacement of the old controllers and encoders.
- Electron Beam Welder (EBW)
    - Completed approval of the work authorizing document for operational and maintenance activities.
    - Performed testing and received the annual certification for a radiation generating device.
    - Revised post-maintenance acceptance test criteria.
    - Scheduled vendor visit review of trouble shooting issues and corrective maintenance
- Surface Preparation
    - Completed installation and assembly of the new unit in the glovebox.

1.3 Issues

- None

1.4 Variance Analysis and Recovery Plan

- None

1.5 BCRs

- None

1.6 Earned Value Metrics

PERFORMANCE S-CURVE
Pit Manufacturing

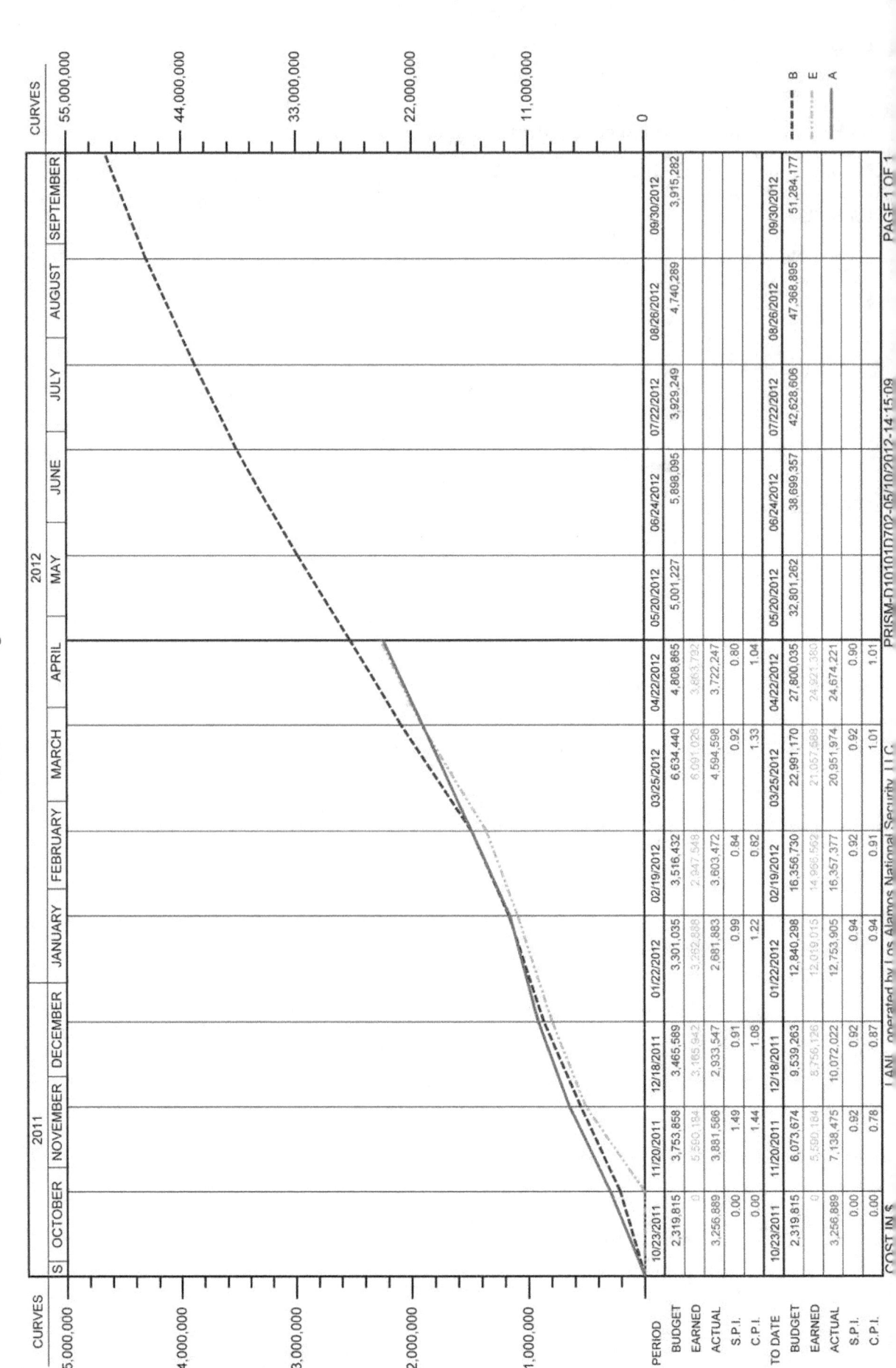

PERIOD		2011			2012								
	S	OCTOBER	NOVEMBER	DECEMBER	JANUARY	FEBRUARY	MARCH	APRIL	MAY	JUNE	JULY	AUGUST	SEPTEMBER
PERIOD		10/23/2011	11/20/2011	12/18/2011	01/22/2012	02/19/2012	03/25/2012	04/22/2012	05/20/2012	06/24/2012	07/22/2012	08/26/2012	09/30/2012
BUDGET		2,319,815	3,753,858	3,465,589	3,301,035	3,516,432	6,634,440	4,808,865	5,001,227	5,898,095	3,929,249	4,740,289	3,915,282
EARNED		0	5,590,184	3,165,942	3,262,888	2,947,548	6,091,026	3,863,792					
ACTUAL		3,256,889	3,881,566	2,933,547	2,681,883	3,603,472	4,594,598	3,722,247					
S.P.I.		0.00	1.49	0.91	0.99	0.84	0.92	0.80					
C.P.I.		0.00	1.44	1.08	1.22	0.82	1.33	1.04					
TO DATE		10/23/2011	11/20/2011	12/18/2011	01/22/2012	02/19/2012	03/25/2012	04/22/2012	05/20/2012	06/24/2012	07/22/2012	08/26/2012	09/30/2012
BUDGET		2,319,815	6,073,674	9,539,263	12,840,298	16,356,730	22,991,170	27,800,035	32,801,262	38,699,357	42,628,606	47,368,895	51,284,177
EARNED		0	5,590,184	8,756,126	12,019,015	14,966,562	21,057,588	24,921,380					
ACTUAL		3,256,889	7,138,475	10,072,022	12,753,905	16,357,377	20,951,974	24,674,221					
S.P.I.		0.00	0.92	0.92	0.94	0.92	0.92	0.90					
C.P.I.		0.00	0.78	0.87	0.94	0.91	1.01	1.01					

COST IN $

CURVES: B E A

2.0 LANL – Experimental Component

4198	GREEN	Perform activities to sustain base pit material processing and fabrication capability through Experimental Component Fabrication

1. Experimental Component Fabrication and Shipment of Gemini / Pollux in accordance with PEP but no later than September 30, 2012

L3 Milestone	Baseline Date	RGYB	Status/Comments
Start Cold Machining	January 10, 2012	BLUE	
Start Hot Machining	February 21, 2012	BLUE	
Casting Complete	February 27, 2012	BLUE	
Complete Castor Sub-Assembly	May 3, 2012	GREEN	Castor sub-assembly will likely be delayed until June per Gemini schedule.
Start Hot Development Assembly	May 7, 2012	GREEN	
Complete Machining Lot 1 – DEV	June 14, 2012	GREEN	
Complete Machining Lot 2 – Pollux	August 1, 2012	GREEN	
Start Pollux Assembly	August 13, 2012	GREEN	
Pollux Complete	September 6, 2012	GREEN	

2.1 Technical Progress

- Completed machining and inspecting the first set of hot parts on Precitech 1. These parts were used to prove-in hot machining routines. One part will be inspected using the CMM and the other will be used to validate the density and radiography fixtures.
- Completed heat treatment of the second (and final) part from Casting Lot 1.
- Received shipping fixtures, mill tooling, and radiographic fixtures.
- Set up mock assembly for shipping "shake test."
- Gluing, assembly, and gas operations procedure continue and are on track.

2.2 Equipment

- Completed refurbishment of the DMU-35 mill controller to provide more stable operation and address the overheating problem.
- Continued constant feed cutting and on machine probing development on cold Precitech.
- Repaired the hot CMM system.

2.3 Issues

- There is concern that mandatory contractor separations will result in the loss of a critical lathe programming resource responsible for on-machine probing and constant feed cutting routines that may be required for Pollux machining. This concern has been elevated to the Associate Director's office.

2.4 Variance Analysis and Recovery Plan

- CPI – Despite requesting cost corrections and recovery plans by the end of April, cost overruns persist. The issue has been elevated to the group and division leads and several recovery options have been proposed for consideration. We expect a plan to be agreed upon in May.
- SPI – The schedule lags due to hardware issues with the CMM and personnel restructuring due to the VSP and Dimensional Inspection group reorganization. Overall, the milestone remains on track, despite the schedule variance.

2.5 BCRs

- None

2.6 Earned Value Metrics

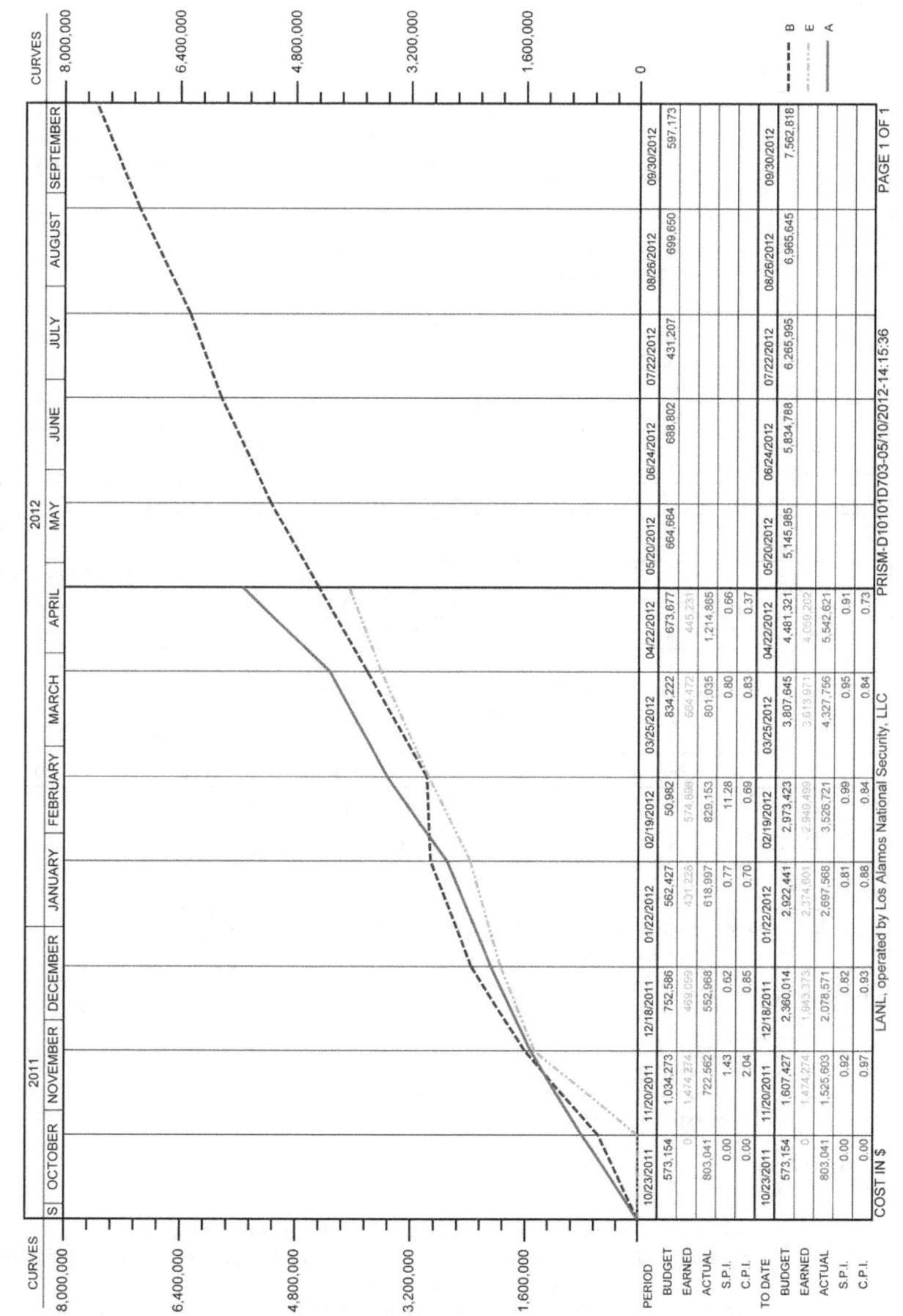

PERFORMANCE S-CURVE
Experimental Component Manufacturing

3.0 LANL - Power Supply

4195	BLUE	Install the Power Supply Assembly Area Equipment and Upgrades per the FY2012 Power Supply PEP

1. Begin construction on the PSAA by March 31, 2012.
2. Start installation of the Analytical Chemistry GBs by June 30, 2012.
3. Start installation of the low voltage E-Beam Welder by June 30, 2012.

L3 Milestone	Baseline Date	RGYB	Status/Comments
Start Construction – Capital (PSAA)	December 15, 2011	BLUE	
Start Construction – Facility (PSAA)	December 15, 2011	BLUE	
GB-1240 - Construction Field Start (Pu Assay)	March 1, 2012	BLUE	
GB-1241 - Construction Field Start (Radiochemistry)	March 9, 2012	BLUE	
Mobilize and Field Start (LVEBW)	June 14, 2012	BLUE	

4197	GREEN	Complete FY 2012 Power Supply material recovery activities IAW the Plutonium Sustainment PEP

1. Dismantle lesser of 480 units or 95% of the Pantex items received by June 30, 2012
2. Recover and store the oxide by Sept 30, 2012

L3 Milestone	Baseline Date	RGYB	Status/Comments
Material Recovery Complete Shipment #1	October 18, 2011	BLUE	
Material Recovery Complete Shipment #2	January 4, 2012	BLUE	
Material Recovery Complete Shipment #3	March 8, 2012	BLUE	
Material Recovery Complete Shipment #4	April 25, 2012	BLUE	
Material Recovery Complete Shipment #5	June 27, 2012	GREEN	
Material Recovery Complete Shipment #6	August 22, 2012	GREEN	

3.1 Technical Progress

- Material Recovery
    - Received shipment #5 (received in April).
    - The Haas Lathe requires field modification to the gas pressure line (limiting valve). A new has been requested.

3.2 Equipment

- Power Supply Assembly Area (PSAA)
    - The project received a Pollution Prevention "Gold" award in the category of Cradle-to-Cradle – Cleanouts. A gold award is presented to the number one project in its pollution prevention category.
    - Construction continues and on track.
    - Completed concrete saw floor cutting and started rebar and new concrete placement (ACB pads).

- o Completed potholing, road milling and the installation of the first (of two) manholes as part of the underground duct-bank installation.
- The Low Voltage Electron Beam (LVEB) Welder
  - o Construction activities started for April.
  - o Vendor on track for acceptance visit planned for 3rd quarter (3 LANL staff to Germany).
- Analytical Chemistry gloveboxes (Pu Assay and Radiochemistry)
  - o The glovebox delivery is on track.
  - o Construction activities started for April.

3.3 Issues

- None

3.4 Variance Analysis and Recovery Plan

- None

3.5 BCRs

- None

3.6 Earned Value Metrics

PERFORMANCE S-CURVE
Reconstitution of Power Supply Manufacturing

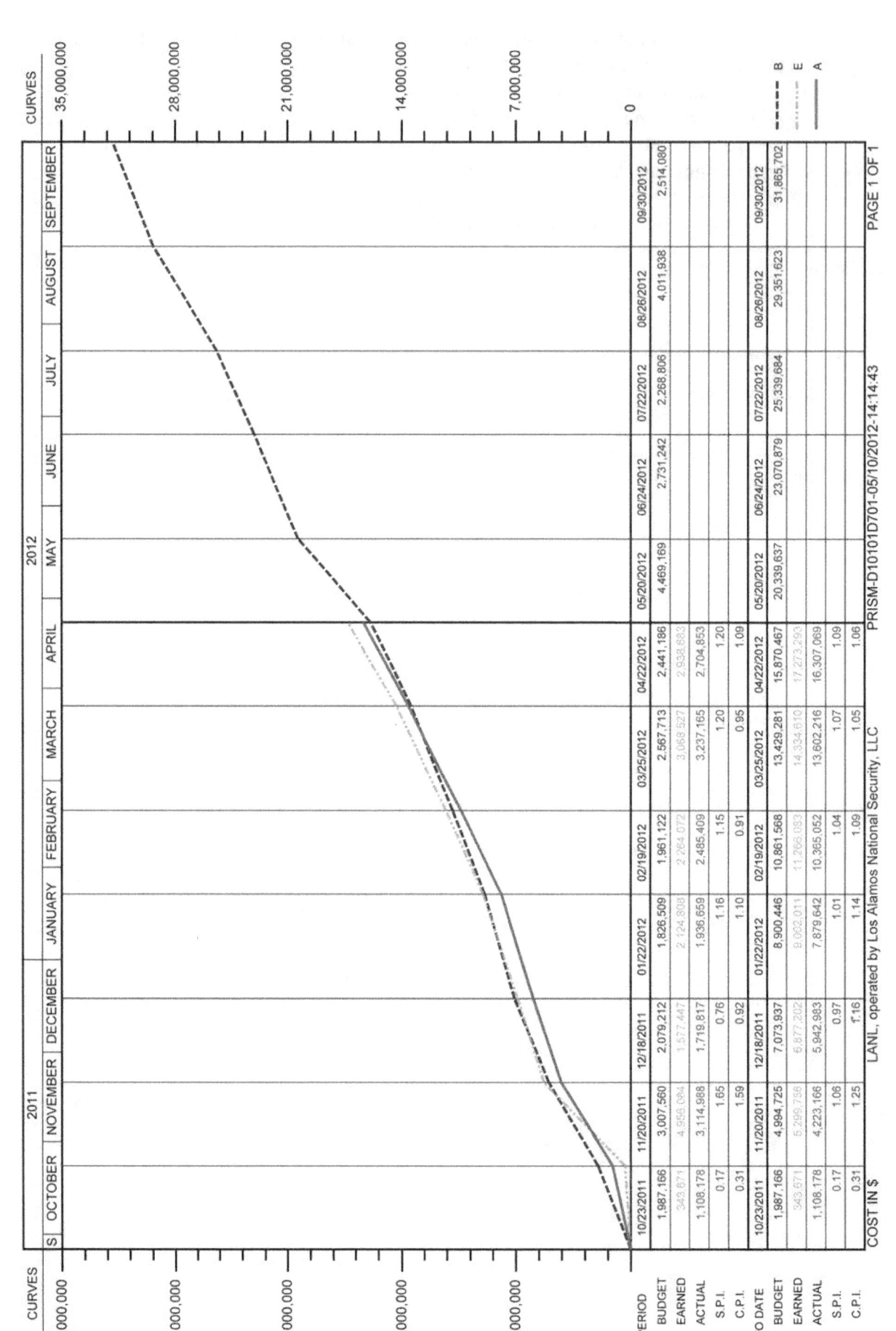

PERIOD		10/23/2011	11/20/2011	12/18/2011	01/22/2012	02/19/2012	03/25/2012	04/22/2012	05/20/2012	06/24/2012	07/22/2012	08/26/2012	09/30/2012
	BUDGET	1,987,166	3,007,560	2,079,212	1,826,509	1,961,122	2,567,713	2,441,186	4,469,169	2,731,242	2,268,806	4,011,938	2,514,080
	EARNED	343,671	4,956,084	1,577,447	2,124,808	2,264,072	3,068,527	2,933,682					
	ACTUAL	1,108,178	3,114,988	1,719,817	1,936,659	2,485,409	3,237,165	2,704,853					
	S.P.I.	0.17	1.65	0.76	1.16	1.15	1.20	1.20					
	C.P.I.	0.31	1.59	0.92	1.10	0.91	0.95	1.09					
TO DATE		10/23/2011	11/20/2011	12/18/2011	01/22/2012	02/19/2012	03/25/2012	04/22/2012	05/20/2012	06/24/2012	07/22/2012	08/26/2012	09/30/2012
	BUDGET	1,987,166	4,994,725	7,073,937	8,900,446	10,861,568	13,429,281	15,870,467	20,339,637	23,070,879	25,339,684	29,351,623	31,865,702
	EARNED	343,671	5,299,756	6,877,202	9,002,011	11,266,083	14,334,610	17,273,293					
	ACTUAL	1,108,178	4,223,166	5,942,983	7,879,642	10,365,052	13,602,216	16,307,069					
	S.P.I.	0.17	1.06	0.97	1.01	1.04	1.07	1.09					
	C.P.I.	0.31	1.25	1.16	1.14	1.09	1.05	1.06					

COST IN $ LANL, operated by Los Alamos National Security, LLC   PRISM-D10101D701-05/10/2012-14:14:43   PAGE 1 OF 1

4.0 Program Management and Support

4.1 Technical Progress

- Analyzed the program impact of the LANL Voluntary Separation Program. The analysis determined an impact of 24.6 FTE, which equates to a budget impact of approximately $3.2 Million Dollars of pull back. Approximately $1.3 Million Dollars is tied to RTBF.
- Delivered Revision 3 of the Plutonium Sustainment Manufacturing Study, which incorporated changes needed due to the release of the FY2013 President's Budget and the delay in the Chemistry and Metallurgy Research Replacement Nuclear Facility (CMRRNF).
- Attended the Pu Sustainment Second Quarter Mid-Year Review providing technical updates across the Plutonium Sustainment portfolio.
- Attended the Pit Production Integrated Project Team meeting in Livermore. Significant effort was focused on reviewing drawing and specification requests, made by LANL, based on preparations for the EDU builds. Both LANL and Kansas City Plant presented on their capabilities for fulfilling the non-nuclear component mission.

4.2 Equipment

- None

4.3 Issues

- None

4.4 Variance Analysis and Recovery Plan

- None

4.5 BCRs

- None

4.6 Earned Value Metrics

PERFORMANCE S-CURVE
Program Office

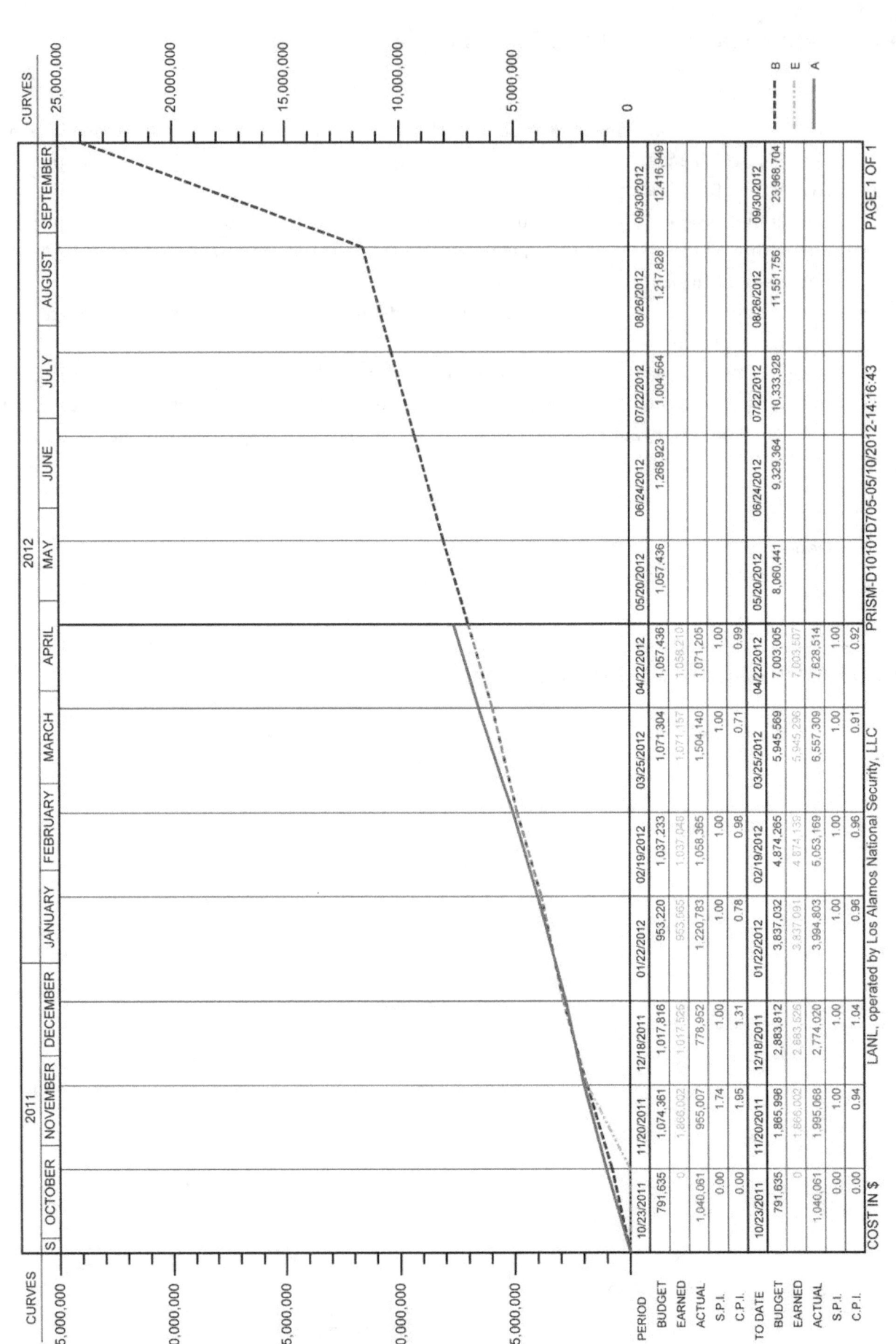

PERIOD		2011			2012								
	S	OCTOBER	NOVEMBER	DECEMBER	JANUARY	FEBRUARY	MARCH	APRIL	MAY	JUNE	JULY	AUGUST	SEPTEMBER
PERIOD		10/23/2011	11/20/2011	12/18/2011	01/22/2012	02/19/2012	03/25/2012	04/22/2012	05/20/2012	06/24/2012	07/22/2012	08/26/2012	09/30/2012
BUDGET		791,635	1,074,361	1,017,816	953,220	1,037,233	1,071,304	1,057,436	1,057,436	1,268,923	1,004,564	1,217,828	12,416,949
EARNED		0	1,866,002	1,017,526	953,565	1,037,046	1,071,157	1,058,210						
ACTUAL		1,040,061	955,007	778,952	1,220,783	1,058,365	1,504,140	1,071,205						
S.P.I.		0.00	1.74	1.00	1.00	1.00	1.00	1.00						
C.P.I.		0.00	1.95	1.31	0.78	0.98	0.71	0.99						
TO DATE		10/23/2011	11/20/2011	12/18/2011	01/22/2012	02/19/2012	03/25/2012	04/22/2012	05/20/2012	06/24/2012	07/22/2012	08/26/2012	09/30/2012
BUDGET		791,635	1,865,996	2,883,812	3,837,032	4,874,265	5,945,569	7,003,005	8,060,441	9,329,364	10,333,928	11,551,756	23,968,704
EARNED		0	1,866,002	2,883,526	3,837,091	4,874,135	5,945,296	7,003,507						
ACTUAL		1,040,061	1,995,068	2,774,020	3,994,803	5,053,169	6,557,309	7,628,514						
S.P.I.		0.00	1.00	1.00	1.00	1.00	1.00	1.00						
C.P.I.		0.00	0.94	1.04	0.96	0.96	0.91	0.92						

COST IN $

CURVES: B (- - - -), E, A

LANL, operated by Los Alamos National Security, LLC   PRISM-D10101D705-05/10/2012-14:16:43   PAGE 1 OF 1

5.0 Facilities, Waste and Institutional Support

5.1 Technical Progress

PF-4 Facility Availability: April 93.43%

- 100 Area: 93.72 %
- 200 Area: 91.67 %
- 300 Area: 94.71 %
- 400 Area: 93.63 %

The PF-4 Facility went to MODE-2 (standby) on April 19, 2012 due to a missed TSR required Surveillance

5.2 Equipment
- None

5.3 Issues

- None

5.4 Variance Analysis and Recovery Plan

- None

5.5 BCRs

- None

5.6 Earned Value Metrics

PERFORMANCE S-CURVE
Facility Infrastructure

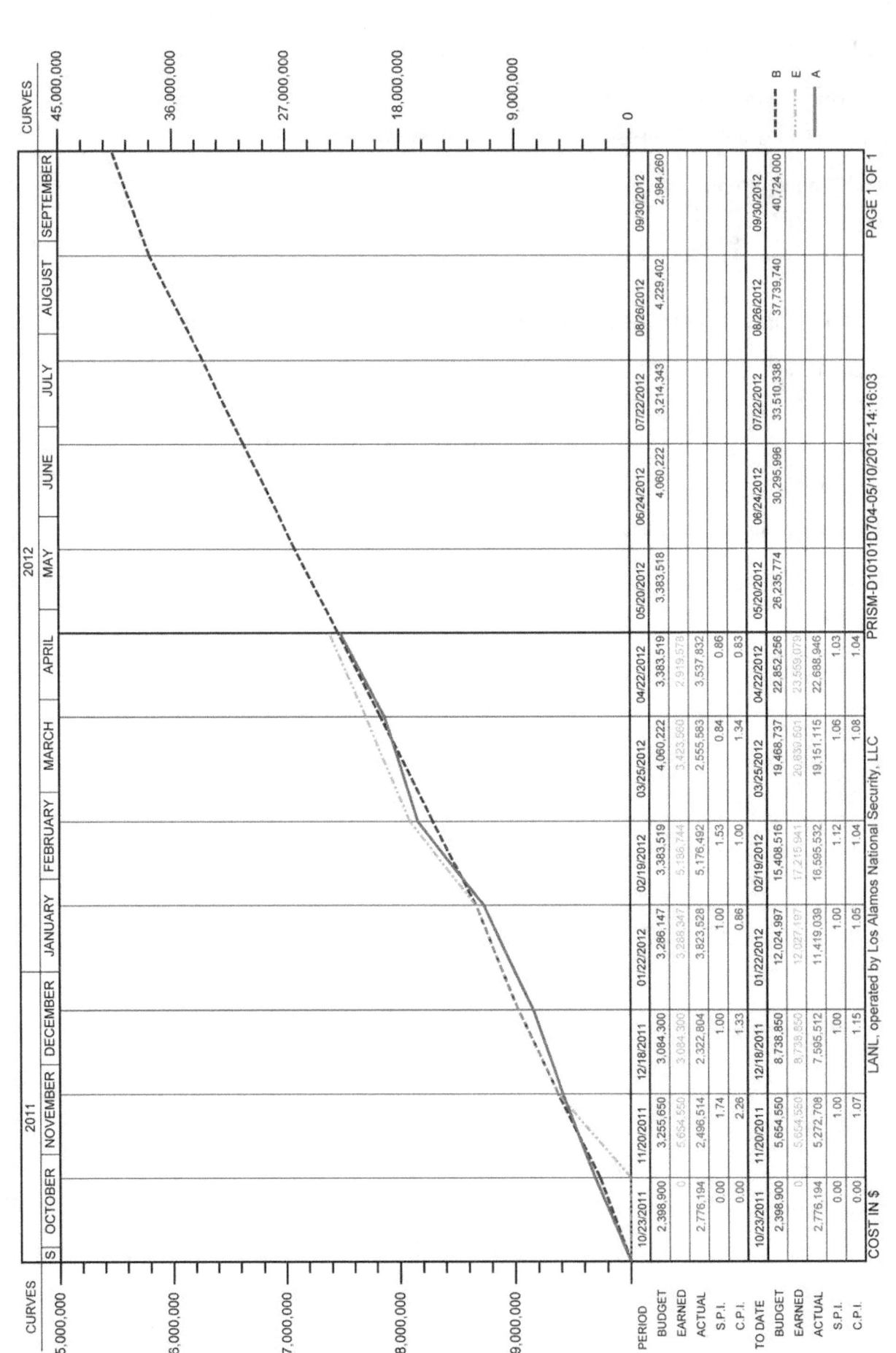

PERIOD	2011			2012								
	OCTOBER	NOVEMBER	DECEMBER	JANUARY	FEBRUARY	MARCH	APRIL	MAY	JUNE	JULY	AUGUST	SEPTEMBER
	10/23/2011	11/20/2011	12/18/2011	01/22/2012	02/19/2012	03/25/2012	04/22/2012	05/20/2012	06/24/2012	07/22/2012	08/26/2012	09/30/2012
BUDGET	2,398,900	3,255,650	3,084,300	3,286,147	3,383,519	4,060,222	3,383,519	3,383,518	4,060,222	3,214,343	4,229,402	2,984,260
EARNED	0	5,654,550	3,084,300	3,288,347	5,136,744	3,423,560	2,919,578					
ACTUAL	2,776,194	2,496,514	2,322,804	3,823,528	5,176,492	2,555,583	3,537,832					
S.P.I.	0.00	1.74	1.00	1.00	1.53	0.84	0.86					
C.P.I.	0.00	2.26	1.33	0.86	1.00	1.34	0.83					
TO DATE	10/23/2011	11/20/2011	12/18/2011	01/22/2012	02/19/2012	03/25/2012	04/22/2012	05/20/2012	06/24/2012	07/22/2012	08/26/2012	09/30/2012
BUDGET	2,398,900	5,654,550	8,738,850	12,024,997	15,408,516	19,468,737	22,852,256	26,235,774	30,295,996	33,510,338	37,739,740	40,724,000
EARNED	0	5,654,550	8,738,850	12,027,197	17,215,941	20,630,501	23,559,079					
ACTUAL	2,776,194	5,272,708	7,595,512	11,419,039	16,595,532	19,151,115	22,688,946					
S.P.I.	0.00	1.00	1.00	1.00	1.12	1.06	1.03					
C.P.I.	0.00	1.07	1.15	1.05	1.04	1.08	1.04					

CURVES: B, E, A

COST IN $

PRISM-D10101D704-05/10/2012-14:16:03

PAGE 1 OF 1

6.0 LLNL

MRT 4196	GREEN	*Perform Activities to Sustain Base Pit Material Processing and Fabrication Capability*
1. Complete FY12 LLNL tasks per PEP and W87 Legacy IPT necessary to support FY2013 delivery of 1 pit EDU 2. Perform initial cold testing of modern foundry 3. Evaluate selected Pit Manufacturing Processes for initial phase of "W87 -Like" fabrication by Sept 30, 2012 4. No later then June 30, 2012, Package and Ship Pu metal to LANL		

L3 Milestone	Baseline Date	RGYB	Status/Comments
LLNL Tasks to supports FY2013 EDU	9/30/2013	GREEN	Design Definition complete for EDU Working chemistry specification
Initial cold testing of modern foundry	9/30/2012	GREEN	Several systems have been cold tested
Evaluate selected Pit Manufacturing Processes for initial phase of "W87 -Like" fabrication	9/30/2012	YELLOW	Slow progress in getting procedures approved. Initial weld testing underway Fixtures delivered
Package and Ship Pu metal to LANL	6/30/2012	GREEN	First shipment completed

6.1 Technical Progress

- Hosted the Pu Sustainment Mid Year Review providing technical updates and tours.

- The Pit Production IPT met on 4/23 and 4/24 in Livermore. A number of topics were discussed including status of LANL transition builds, plans and criteria for EDU, and the non-nuclear component mission assignment. Both LANL and Kansas City Plant presented on their capabilities for fulfilling this mission. Additionally, during this meeting, significant effort was focused on reviewing drawing and specification requests, made by LANL, based on their current preparations for transition and EDU builds. LLNL will review the requests and issue drawing and specification updates where appropriate. LLNL will provide an update on a QC-1 maintenance plan for the Sheffield at the next Pu Sustainment Review.

- Tube laser brazing tests continue.

- Fixtures for the Moore T-Base lathe #1 have been completed and sent to LANL.

- The Sheffield inspection procedure is in USQ review and should be approved within a week.

- Procedure now in place for LANL to complete repackaging and confirmatory NDA of nuclear parts so that LLNL can begin datum machining parts on Hardinge lathe.

- Work continues on building Sheffield inspection fixture and mock shell.

- Metal for LANL. All processing has been completed. 33 items have been made; 20 sent to LANL. Remainder will be sent to meet milestone.

- Hot press operations are continuing.

6.2 Equipment

- Modern Foundry :
  - o Continued cold testing the transport system and verifying TC and control system wiring. Cold testing primarily involves software programming and the associated program nuances and error identification. TC and control system wiring involves verification of signal generation to control computer interface.

6.3 Issues

- Need LANL to complete their use of Moore T-Base lathe #1 so LLNL can program and perform final chuck machining and machine programming.

- Need LANL to complete NDA of nuclear parts so that datum machining can begin.

6.4 Variance Analysis and Recovery Plan

- None

6.5 BCRs

- None

6.6 Earned Value Metrics

PERFORMANCE S-CURVE
LLNL

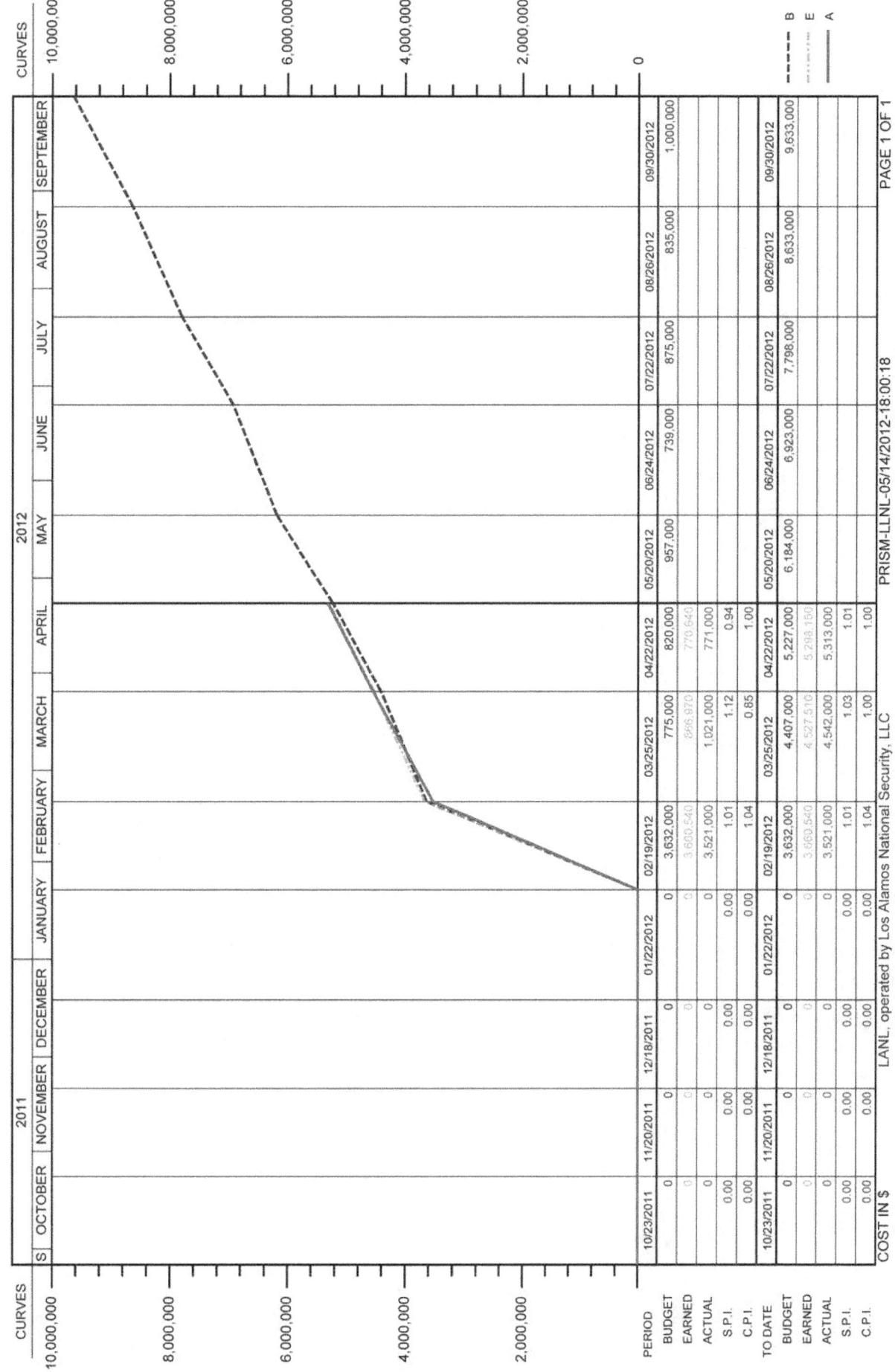

PERIOD	S	OCTOBER	NOVEMBER	DECEMBER	JANUARY	FEBRUARY	MARCH	APRIL	MAY	JUNE	JULY	AUGUST	SEPTEMBER
		10/23/2011	11/20/2011	12/18/2011	01/22/2012	02/19/2012	03/25/2012	04/22/2012	05/20/2012	06/24/2012	07/22/2012	08/26/2012	09/30/2012
BUDGET		0	0	0	0	3,632,000	775,000	820,000	957,000	739,000	875,000	835,000	1,000,000
EARNED						3,660,540	896,970	770,640					
ACTUAL		0	0	0	0	3,521,000	1,021,000	771,000					
S.P.I.		0.00	0.00	0.00	0.00	1.01	1.12	0.94					
C.P.I.		0.00	0.00	0.00	0.00	1.04	0.85	1.00					
TO DATE		10/23/2011	11/20/2011	12/18/2011	01/22/2012	02/19/2012	03/25/2012	04/22/2012	05/20/2012	06/24/2012	07/22/2012	08/26/2012	09/30/2012
BUDGET		0	0	0	0	3,632,000	4,407,000	5,227,000	6,184,000	6,923,000	7,798,000	8,633,000	9,633,000
EARNED						3,660,540	4,527,510	5,298,150					
ACTUAL		0	0	0	0	3,521,000	4,542,000	5,313,000					
S.P.I.		0.00	0.00	0.00	0.00	1.01	1.03	1.01					
C.P.I.		0.00	0.00	0.00	0.00	1.04	1.00	1.00					

COST IN $

LANL, operated by Los Alamos National Security, LLC

PRISM-LLNL-05/14/2012-18:00:18

PAGE 1 OF 1